THE COMPLETE SPACE BUFF'S

BUCKET LIST

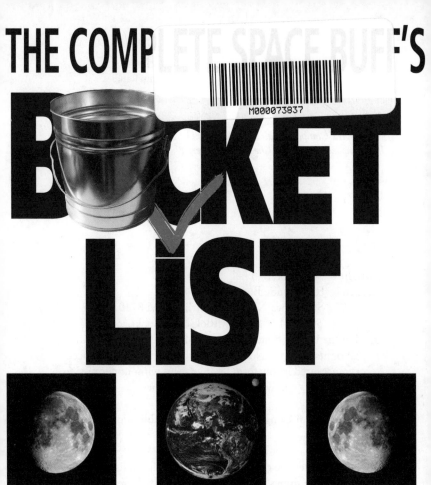

100 Space Things to Do Before You Die
by **Loretta Hall**
with a few words
by **Mike Mullane & Dr. David Livingston**

Río Grande Books
Los Ranchos, NM

Río Grande Books

The Bucket List Book Series

The Complete Cowboy Bucket List by Slim Randles
The Ultimate Hot Air Balloon Bucket List by Barbe Awalt
The Basic New Mexico Bucket List by Barbe Awalt
The Complete Space Buff's Bucket List by Loretta Hall (2016)
The Complete Crab Lovers' Bucket List by Barbe Awalt (2016)
The Complete Green Chile Cheesburger Lovers' Bucket List by Barbe Awalt (2016)

Copyright © 2016 Loretta Hall
Published by Río Grande Books
925 Salamanca NW, Los Ranchos, NM 87107-5647
505-344-9382 www.RioGrandeBooks.com
Printed in the United States of America Book Design: Paul Rhetts

Library of Congress Cataloging-in-Publication Data

Names: Hall, Loretta.
Title: The complete space buff's bucket list :
100 space things to do before you die / by Loretta Hall.
Description: Los Ranchos, NM : Rio Grande Books, [2015]
| Series: The bucket list book series

Identifiers: LCCN 2015031298
| ISBN 9781943681013 (pbk. : alk. paper)

Subjects: LCSH: Astronautics--Popular works.
Classification: LCC TL793 .H34216 2015 | DDC 629.4--dc23
LC record available at http://lccn.loc.gov/2015031298

*Front cover: The Big Dipper, seen from the International Space Station.
Courtesy of NASA*

CONTENTS

Acknowledgements

Reflecting on the spectrum of places to visit and things to do that comprise this book, I realize I am deeply grateful to all of the space aficionados who provide these opportunities. I appreciate all of the people I have met in museums and shops, at space conferences, and in email correspondence. Their degree of lightheartedness varies, but the enthusiasm is always there.

I want to thank the people who provided images for this book and permission to use them. The photographs and graphics they supplied truly enhance this book. As you enjoy the following pages, please notice the businesses credited in the captions and visit their websites.

I am particularly grateful to Mike Mullane and David Livingston, who agreed to preview this book's content and comment on it. I also appreciate the suggestions my fellow space buff Ted Spitzmiller made for the "Near Misses" page.

Thanks to my publishers, Barbe Awalt and Paul Rhetts at Rio Grande Books, who are producing a series of Bucket List books on a broad range of topics. I have had great fun searching out—and in some cases, dreaming up—the 100+ items in *The Complete Space Buff's Bucket List*. I have discovered many new things to do and see.

Thanks also to my husband, Jerry, for his enthusiastic support of my projects, in the creative aspect of discussing ideas and in the practical aspect of proofreading. If any errors remain in this book, they are probably in things I changed after he read it.

Test launch of Blue Origin's New Shepard. Courtesy of Blue Origin

Experiencing the Final Frontier

What a wonderful book! As a space shuttle astronaut, I was blessed to experience space from a very unique perspective. But my passion didn't end at the last "Wheel Stop" call. As it is for many of us, there's a gene in my Being that continues to command me to look up and wonder. With *The Complete Space Buff's Bucket List* I now have a new resource to map my way back into space. I can roam through distant galaxies via the Hubble Space Telescope (Bucket List item 65). Item 59 will allow me to stroll on Mars and our other planetary neighbors. Reading Carl Sagan's *Cosmos* warp-speeds me back to the beginning of time (item 100). But there are so many other waypoints on the journey and Loretta Hall's book has succinctly provided the coordinates for those stops. See you in space! (item 1) — Mike Mullane, NASA Astronaut (Ret.), Author of *Riding Rockets: The Outrageous Tales of a Space Shuttle Astronaut*

Loretta's *The Complete Space Buff's Bucket List: 100 Space Things to Do before You Die* is an excellent list of exciting and worthwhile things to do in your lifetime, either alone or with your family or friends. While many will be familiar with some of the items on the list, many of the items are not so familiar. So her "bucket list" can easily serve as a valuable tool for us all as we engage in different, new, and exciting events, visit places, and strive to learn new things in our lifetimes. I recommend this book and the magical 100 items on the list for everyone regardless of age, space experience, or knowledge. Her list is universal. Perhaps I will get to meet some of you as we take on visiting these places and doing these things in the coming years. I look forward to it. — Dr. David Livingston, Host of *The Space Show*

Artist's concept of World View flight. Courtesy of World View

Being a Space Buff

I have been a space buff since 1959, when NASA [the National Aeronautics and Space Administration] introduced the Mercury Seven astronauts to the world. These were the first Americans chosen to fly into space. Over the years, I have watched amazing feats and tragic disasters. Men bringing rocks back from the Moon. Men and women living in Earth orbit for months at a time. Three fatal NASA disasters and some near misses. Unmanned probes collecting atmospheric data and photographing planets throughout our solar system … and beyond.

For the true space enthusiast, reading news accounts of such adventures is not enough. We need details. We want to be involved. We want to participate.

The bucket list I have put together in this book is my vision of the most intriguing and satisfying activities to complete my fascination with space exploration. Some are simple, even silly. Destinations are located in the United States, but similar or analogous destinations can be found abroad. Some are time consuming or expensive. In some cases, I have given prices; they may change by the time you read this book, but I wanted to give you an idea of the order of magnitude for planning and comparison. Together, the 100 items on this list encompass a broad range of intellectual and physical activities that involve all five senses.

Most items on this Bucket List include URLs for websites associated with that item. Remember that URLs are not case sensitive, so don't worry about getting all the right letters capitalized. I'm sorry that some of them are long and complex, but I didn't make them up.

Now, start reading and join me in living our dream!

Carl Sagan's books. Photo by Loretta Hall

The List:

100. Read Carl Sagan's *Cosmos* or *Contact*

Astronomer Carl Sagan participated in NASA's Mariner, Viking, and Voyager planetary exploration missions. Sagan became a celebrity scientist through his books and television series. Written in 1980, *Cosmos* is a beautifully illustrated history of the universe and man's exploration of it. *Contact* is a fictional account of interaction with intelligent extraterrestrial life.

Other books by Carl Sagan:

Billions & Billions: Thoughts on Life and Death at the Brink of the Millennium

Broca's Brain: Reflections on the Romance of Science

Cosmic Connection: An Extraterrestrial Perspective

The Demon-Haunted World: Science as a Candle in the Dark

The Dragons of Eden: Speculations on the Evolution of Human Intelligence

Pale Blue Dot: A Vision of the Human Future in Space

Shadows of Forgotten Ancestors

The Varieties of Scientific Experience: A Personal View of the Search for God

99. Tune in to NASA TV

Find out what is happening in NASA missions, programs, activities, and science developments. Watch daily broadcasts from the International Space Station (ISS) and vicariously experience living in orbit and taking spacewalks. — https://www.nasa.gov/multimedia/nasatv/#.VOUvji7lyTZ

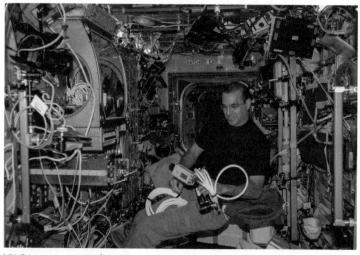

NASA astronaut Rick Mastracchio working aboard the ISS. Courtesy of NASA.

98. Visit a planetarium

Learn about stars, constellations, planets, dark matter, and more. Dramatic movies of flights through space surround you when projected on the dome's interior. The Adler Planetarium in Chicago, the Burke Baker Planetarium in Houston, and the Hayden Planetarium at the American Museum of Natural History in New York City are among the best. For a planetarium near you, do an Internet search on the word *planetarium* and the name of your state. You might be surprised how close one is!

Courtesy of SUNY Oswego. Photo by Jim Russell

97. Identify an astronaut you are interested in and read his/her memoir

Many astronauts have written books about their experiences in space. For example, America's first astronauts, the Mercury Seven, compiled their recollections in *We Seven: By the Astronauts Themselves*. Michael Collins, command module pilot for Apollo 11, wrote *Carrying the Fire: An Astronaut's Journeys*. Mission specialist Mike Mullane wrote *Riding Rockets: The Outrageous Tales of a Space Shuttle Astronaut*. Sally Ride, America's first female astronaut, wrote *To Space and Back*. And there are many more to choose from.

Astronauts. Courtesy of NASA

96. Attend a space conference

Connect with other space enthusiasts and find out what is happening in governmental and commercial space ventures. Hobnob with those in the know. Annual possibilities include the Space Frontier Foundation's NewSpace Conference in Silicon Valley, California, the National Space Society's International Space Development Conference, and the Mars Society Convention. Combine this Bucket List item with items 30-25 by attending the International Symposium for Personal and Commercial Spaceflight in Las Cruces, New Mexico, in early October every year and taking side trips to other southern New Mexico space sites. — https://newspace. spacefrontier.org, http://isdc.nss.org, http://www.marssociety. org, http://www.ispcs.com

International Symposium for Personal and Commercial Spaceflight. Photo by Loretta Hall

95. Visit Tomorrowland

In 1955, it was one of the original lands in Disneyland Park in Anaheim, California, and it has been updated over the years. Since 1971, Tomorrowland has also been part of the Magic Kingdom Park at the Walt Disney World Resort near Orlando, Florida. Both feature the popular Space Mountain adventure, but other attractions differ between the two parks. — https://disneyland.disney.go.com/au/disneyland/tomorrowland, http://wdwinfo.com/wdwinfo/guides/magickingdom/tomorrow.htm

Tomorrowland sign at Disneyland. Public domain

94. Experience Mission: SPACE in Epcot

A visit to Tomorrowland in Magic Kingdom Park at the Walt Disney World Resort in Florida is fun, but you can enhance your experience by visiting another park, Epcot. Epcot offers the Mission: SPACE thrill ride, a simulated trip to Mars. Choose the Orange Team training level for a centrifuge ride, or pick the Green Team option of a non-spinning motion simulator if you might be prone to motion sickness. During the flight you'll have an active role to play, so pay attention during the training! — https://disneyworld.disney.go.com/attractions/epcot/mission-space

Mission: SPACE sign at Epcot. Creative Commons license, photo by Michael Gray

93. The Mercury 13: Gender discrimination or program expediency?

Following the physical exams of the Mercury astronaut candidates in 1959, Dr. Randy Lovelace decided to perform the same series of exhaustive tests on highly accomplished female airplane pilots. Thirteen of the women were found to be as physically qualified as the Mercury 7 male astronauts. The US government denied them the opportunity for further astronaut candidate testing. Some saw this as gender discrimination, while others said it would cause a delay in the tight time schedule President John F. Kennedy set for landing an American on the Moon. Read both sides of the debate and decide for yourself. One side is explained in *Right Stuff, Wrong Sex* by Margaret Weitekamp, and a rebuttal appears at http://thespacereview.com/article/869/1. Find out about the individual ladies at http://mercury13.com/ladies.htm.

Seven members of the Mercury 13. Courtesy of NASA

92. Sit back and take a virtual tour of Goddard Space Flight Center

Located in Greenbelt, Maryland, Goddard Space Flight Center was the first of NASA's spaceflight complexes. It is open for tours only to school, community, and cultural groups, but its website offers an interesting virtual tour that is described as an interactive experience for you to "learn about the science, people, and processes that transform a NASA spacecraft from an idea into a reality." — http://www.nasa.gov/externalflash/goddardVT

Virtual tour screen shot. Courtesy of Goddard Space Flight Center

91. Get a preview of the Museum of Science Fiction

Many astronauts as well as scientists and engineers in space programs were inspired by reading science fiction. Among them are Robert Goddard, Stephen Hawking, Yuri Gagarin, Peter Diamandis, Eileen Collins, and Wernher von Braun (who even wrote a sci-fi novel to popularize space exploration). If you don't know who all of those people were/ are, look them up! At this writing, the Museum of Science Fiction is still in the planning stages, but you can see glimpses of what is planned at http://www.museumofsciencefiction. org/proposed-galleries. A Preview Museum will tour major US cities before becoming a part of the permanent museum when it is built in Washington, DC.

Museum of Science Fiction logo. Courtesy of Museum of Science Fiction.

90. Visit the Smithsonian National Air & Space Museum

See artifacts ranging from the Wright Brothers' airplane to lunar landers. The main museum is located on the National Mall in Washington, DC, and admission is free. There's nothing like seeing the exhibits in person, appreciating their sizes, and observing their details and textures. If you must wait for a chance to visit the District of Columbia, start by exploring the museum's online exhibits. — http://airandspace.si.edu/exhibitions/online

V-2 (larger) and WAC Corporal missiles. Photo by Loretta Hall

89. Visit the Steven F. Udvar-Hazy Center

While you're in the nation's capital, take a side trip to nearby Chantilly, Virginia, and visit this companion facility of the Smithsonian National Air & Space Museum. Walk among numerous historic airplanes, see the space shuttle *Discovery*, ride flight simulators, and watch preservationists restore historic air and space vehicles and parts. — http://airandspace.si.edu/visit/udvar-hazy-center

Space Shuttle Discovery. Photo by Loretta Hall

88. Watch *Gravity* and *The Martian*

The 2013 science fiction film *Gravity* may not be completely accurate from a technical standpoint, but it was very popular with the public. It also garnered praise for the technical accuracy of some parts of the film. See a discussion of those aspects at http://spacesafetymagazine.com/space-debris/kessler-syndrome/expert-views-gravity-great-movie-bad-science-accurate-realism-three. *The Martian*, a 2015 film, received NASA's approval for its plausibility. See a discussion at https://www.nasa.gov/feature/nine-real-nasa-technologies-in-the-martian.

NASA astronaut on spacewalk. Courtesy of NASA.

87. Own a piece of the extraterrestrial universe

Buy a meteorite. Meteorites For Sale offers stone, iron, and stony-iron meteorites, along with tektites, impactites, and moldavites (look them up!). If you really want to rock, buy a meteorite that came from the Moon or Mars. Meteorite prices start at $10 and go sky high. Moon and Martian meteorites start at $40-$50, but quantities are limited. — https://www.meteorites-for-sale.com

Lunar meteorite in display case. Courtesy of NASA.

86. Learn to hunt for meteorites

Cut out the middleman and enjoy the adventure of finding meteorites yourself. You'll need some equipment, some knowledge of what to look for, and an idea of where to search. There is a wealth of great information at The Meteorite Exchange website. You will not only learn how to become a meteorite hunter, but you'll discover how fascinating these space rocks are. Meteorites can fall to Earth anywhere, but experts recommend dry lake beds in arid places like Arizona and New Mexico for starters because they are unobstructed and the light-colored soil contrasts with dark meteorites. — http://www.meteorite.com/finding-meteorites

Meteorite in Mojave Desert. Creative Commons License. photo by Meteoritekid

85. Visit the Barringer Meteorite Crater

The meteorites you find or buy are small specimens. Blow your meteoritic mind by visiting the best preserved impact crater on this planet. It is located 5 miles south of Interstate 40, about 35 miles east of Flagstaff, Arizona. The circular crater is nearly a mile wide and more than 550 feet deep. Confused about the name—Meteor Crater or Meteorite Crater? When a meteoroid (a small rocky or metallic body travelling through space) enters the Earth's atmosphere, friction and chemical reactions cause it to become a fireball (meteor). If it survives that "trial by fire" and lands on the ground, it is a meteorite. Now you know. — http://barringercrater.com, http://meteorcrater.com

Barringer Meteorite Crater. Courtesy of USGS

84. Tour Biosphere 2

What is it like to live in a completely enclosed, self-sustaining habitat? To get an idea of what colonists on the Moon or Mars will experience, visit Biosphere 2. Beginning in 1991, eight volunteers were sealed in the facility for two years. It is now owned by the University of Arizona and is used as a laboratory for studying Earth systems. It is located just north of Tucson, 260 miles south of Flagstaff (see Bucket List item 85). — http://b2science.org

Biosphere 2. Creative Commons License. photo by CGP Grey

83. Parachute with a partner

Except for the space shuttles, returning spacecraft have employed parachutes to slow them for landing. Find out what it feels like to have a parachute deploy and begin to slow your fall. Tandem skydives, in which you are harnessed to an experienced skydiver, are available at many locations). Allow four and a half hours for the orientation session, flight, freefall (40 to 60 seconds), and parachute ride (5 to 8 minutes). — http://www.incredible-adventures.com/halo-jumping.html, http://www.xperiencedays.com/Skydiving_Experiences.html

Tandem parachute jump. Courtesy of US Air Force. photo by Staff Sgt Megan Crusher

82. Skydive indoors

Not too sure you want to step out of a perfectly good airplane two miles above the ground? Try a simulated skydive in a vertical wind tunnel instead (or as a prelude to the real thing). No parachute is involved; you just experience the sensation of free fall for 60 seconds, supported by a column of air rushing upward through the chamber. Allow a little over an hour for orientation, suiting up, and flying. Facilities are located in many cities around the United States; check the website for locations. — http://www.skyventure.com

Courtesy of iFLY Indoor Skydiving/Skyventure

81. Watch *2001: A Space Odyssey*

The 1968 movie is strangely compelling, but cryptic. Watch it again. And again. Until you understand the ending. If you still don't get it, see http://www.kubrick2001.com.

Scene from 2001: A Space Odyssey. *Courtesy of Warner Home Video*

80. Visit the Museum of Flight

Go to Seattle and enjoy the aircraft and spacecraft exhibits. Be sure to cross the street and visit the Charles Simonyi Space Gallery and its exhibits depicting the first fifty years of spaceflight. For a special treat, sign up for a thirty-minute tour of a space shuttle crew compartment and explore both levels of the actual space shuttle astronaut trainer that was transferred to the museum from Johnson Space Center after the shuttle program ended. — http://www.museumofflight. org

Space shuttle trainer at the Museum of Flight. Photo by Loretta Hall

79. Do a lunar roll call

Learn the names of the twenty-seven men who have traveled to the Moon and back. They flew on the Apollo 8 and 10-17 missions. Who were the twelve who explored the lunar surface? And don't forget the other Apollo missions. Find out who flew on the Apollo 7 and 9 Earth-orbit flights and why their less-glamorous missions were important. — https://www.nasa.gov/mission_pages/apollo/missions/index.html#.VOpnLi7lyTa

Apollo program insignia. Courtesy of NASA

78. Satisfy the requirements for the BSA Space Exploration merit badge

This is a multi-step project, and you'll have to learn some science—but that's good for you. You won't have a counselor to supervise your progress (unless you actually are a Boy Scout), so you'll be on the honor system—which is appropriate for a BSA endeavor, after all. See the Space Exploration merit badge requirements at http://usscouts.org/usscouts/mb/mb107.asp.

Boy Scouts of America Space Exploration Merit Badge

77. Dress like an astronaut

For Halloween or another costume event, choose a simple jumpsuit or the full space suit and helmet gear. For instructions on making your own space suit costume, see http://www.artistshelpingchildren.org/howtomakeastronautcostumeshalloweencraftskids.html.

While you are at it, learn about the components of a real extravehicular mobility suit that can sustain a spacewalking astronaut for as long as eight hours. — https://www.nasa.gov/audience/foreducators/spacesuits/home/clickable_suit.html#.ViWGNStyzhV,

http://i.space.com/images/i/000/030/895/i02/space-suit-components-130716c-02.jpg?1374016385

For everyday wear, get a flight jacket from NASA. You can even choose a style with a personalized nametag. Caution: strangers may ask for your autograph. — http://www.thespacestore.com/nasa-clothing/nasa-flight-jackets

Apollo 11 astronaut Buzz Aldrin in his space suit. Courtesy of NASA/KSC

76. Discover the wonders of duct tape

Duct tape has been a staple on Gemini, Apollo, space shuttle, and International Space Station missions. It allowed the Apollo 13 crew to adapt the command module's square air purification cartridge to fit the circular opening in the Lunar Module environmental system. Apollo 16 and 17 lunar landers even took it to the surface of the Moon. It did not stick very well to surfaces covered with the fine lunar dust, but Apollo 17 astronauts were able to successfully patch a damaged fender on the lunar rover with duct tape and a folded map. See a list of duct tape applications on Apollo 9 through Apollo 17 at http://www.workingonthemoon.com/WOTM-DuctTape.html. Keep a roll of duct tape handy and respect it as never before.

Duct tape repair on Apollo 13. Courtesy of NASA

75. Tour the Air Force Space and Missile Museum

Go to the home of NASA's manned space launches, Cape Canaveral, Florida. At the Air Force Space and Missile Museum, you can see Launch Complexes 5/6 and 26, which were used in preliminary and suborbital manned flights of the Mercury Program. Walk through a launch blockhouse. See memorabilia of *I Dream of Jeannie*, a 1965-1970 television comedy show about a NASA astronaut and his genie, Jeannie, who lived in an exotic bottle. The museum's website also offers an extensive virtual tour so you can see the sites and exhibits from the comfort of home. — http://afspacemuseum.org

Inside Blockhouse 26. Courtesy of Bubba73 at English Wikipedia

74. Visit the National Museum of the US Air Force

Another relevant Air Force museum is located at Wright-Patterson Air Force Base in Dayton, Ohio. In the Missile and Space Gallery, you will see spacecraft, launch vehicles, and other exhibits from the Mercury, Gemini, and Apollo programs. You will see gondolas used in the Manhigh high-altitude balloon flights and Joe Kittinger's ultrahigh-altitude parachute jumps, which helped develop equipment and procedures for manned spaceflight. Walk through a space shuttle payload bay exhibit. This museum also offers a virtual tour. Virtual tours are great for previews and for people who are unable to visit the physical museum, but try to go in person if you can. The experience is much richer. — http://www.nationalmuseum.af.mil

Space shuttle payload bay exhibit at the National Museum of the USAF. Courtesy of US Air Force

73. Watch the first season of the *Star Trek* television series

This popular series, which debuted in 1966, helped make space travel seem possible. Some of the technologies its characters used are now becoming real. It was also hailed as presenting a positive vision of multicultural cooperation, including the first interracial kiss shown on television (in Episode 10 of Season 3). Technically, it was the second, as Sammy Davis, Jr., gave Nancy Sinatra a quick peck on the cheek on a TV special several months earlier. However, the Star Trek version was lip-to-lip between Captain Kirk and Lieutenant Uhuru.

Mr. Spock and Captain Kirk. Courtesy of NBC Television

72. Attend a *Star Trek* or *Star Wars* convention

Let your hair down and go in costume. Discover what it is like to mingle with creatures from other galaxies. Meet actors, hear sci-fi speakers, and buy memorabilia. Live long and prosper! — http://www.treknews.net/category/convention, http://www.starwars.com/events/conventions.

Costumed attendees at a Star Wars *event. Photo by Loretta Hall*

71. Learn to speak Klingon

Seriously. You can learn to speak the language of the warrior species from Star Trek. Start with an introduction from the Klingon Language Institute (http://www.kli.org). Go higher tech with a Klingon language app for your iPhone (http://www.iphonefaq.org/archives/97728) or Android phone (https://play.google.com/store/apps/details?id=org.tlhInganHol.android.klingonassistant). Unfortunately, stories of a Rosetta Stone computer-based course in Klingon proved to be an April Fool's Day prank. Instead, you can pick up a copy of *How to Speak Klingon: Essential Phrases for the Intergalactic Traveler* by Ben Grossblatt (Chronicle Books, 2013).

Klingon is a language. Photo by Loretta Hall

70. Visit the US Space and Rocket Center

Located near the Marshall Space Flight Center in Huntsville, Alabama, this museum displays two versions of Saturn rockets. Standing vertically outdoors is a full-scale Saturn I rocket, demonstrating its impressive size. Indoors, hanging horizontally from the ceiling is the Saturn V dynamic test vehicle. Only two other locations have Saturn V rockets on display: Johnson Space Center in Houston (see Bucket List item 37) and Kennedy Space Center in Cape Canaveral (see Bucket List item 20). Other exhibits include the Apollo 16 command module and an engineering mockup of Skylab. — http://rocketcenter.com

U.S. Space and Rocket Center. Courtesy of Space Camp®

69. Make a date with Dr. Space

The Space Show radio program broadcasts live over the Internet to listeners in more than fifty countries. Host Dr. David Livingston interviews prominent guests several times a week. Email the host at DrSpace@TheSpaceShow.com to sign up for a free weekly newsletter that highlights the upcoming week's lineup and broadcast times. You can listen to any previous show free online. Scroll through episodes chronologically at the show's website, or do an Internet search for episodes that discussed a particular topic (for example, <"the space show" "new horizons">). When a planned website upgrade is completed, all episodes will be completely searchable. — http://thespaceshow.com

Logo. Courtesy of The Space Show

68. Think you know what Vanguard was?

Think again. Sure, it was NASA's first rocket designed to launch satellites into Earth orbit. Sure, it referred to three Vanguard satellites launched in the late 1950s—the second, third, and fourth successfully launched American satellites, all of which are still in orbit. But what about the USNS Vanguard, an instrumentation ship used as a sea-based tracking and communication station for Apollo Moon missions, the Apollo-Soyuz program, and the Skylab program? — http://navsource.org/archives/09/53/5319.htm.

USNS Vanguard. Courtesy of U. S. Navy

67. Make a playlist of space music

Begin with the popular instrumental "Telstar." The haunting melody was inspired by the July 1962 launch of the first Telstar communications satellite. The original 1962 hit recording by the Tornados featured the eerie sounds of a theremin, an electronic instrument played by waving hands above it. The 1963 version by the Ventures imitated the sound using an electric guitar. Another must for your playlist is the "Sunrise" fanfare from "Also Sprach Zarathustra," which was the opening theme from *2001: A Space Odyssey*. Theme songs from your other favorite space adventure movies are good candidates too. Visit http://setiathome.berkeley.edu/misc.php for some tunes inspired by SETI@home (see Bucket List item 4).

Carolina Eyck playing a theremin. Creative Commons License. photo by Julius Kaiser

66. Visit an observatory

Photographs in astronomy books and space telescope websites are beautiful, but nothing beats a personal experience. Look through a telescope to examine the Moon or the Sun. Extend your visit past dusk and get a close-up view of our galaxy. For an extra treat, attend a star party, where extra telescopes are set up, astronomers give talks, and amateur astronomers are happy to discuss their equipment and experiences with you. The Griffith Park Observatory in Los Angeles, California, and the Apollo Observatory in Dayton, Ohio, are among the most popular. Or find one near you at http://go-astronomy.com/observatories.htm.

The George Observatory. Courtesy of the Houston Museum of Natural Science

65. Gaze through the Hubble space telescope

For a closer look at the universe, examine images from the Hubble space telescope. At http://hubblesite.org, you will find single images and videos. Stunning (but low-resolution) images and videos are also available at https://www.flickr.com/photos/gsfc/sets/72157623410904809. Read articles about the discoveries made by the Hubble space telescope at https://www.nasa.gov/mission_pages/hubble/main/#.VND-dbS7lyTY.

Hubble Space Telescope. Courtesy of NASA

64. Visit the Kansas Cosmosphere and Space Center

Some impressive space-related destinations are a bit off the beaten path. The Cosmosphere and Space Center is in Hutchinson, about 50 miles northwest of Wichita, in south-central Kansas. But it is well worth the trip. It claims "a US space artifact collection second only to the Smithsonian National Air and Space Museum and the largest collection of Russian space artifacts outside of Moscow." Exhibits include the actual Apollo 13 command module, the Liberty Bell 7 Mercury capsule, and a flown Vostok spacecraft. Get a preview of the other treats that await you at http://cosmo.org/mu_space.htm.

Apollo 13 command module on display at the Kansas Cosmosphere. Public domain

63. Eat at a Space Aliens Grill & Bar

Speaking of "off the beaten path" destinations, you will have to visit Albertville, Minnesota, or Bismark or Fargo, North Dakota, to dine at a Space Aliens Grill & Bar (http://spacealiens.com). Menu items include Martian Munchies, Spaceship Supreme Pizza, and a Solar Eclipse. Don't think this is just a gimmicky dive; they boast that their Bar-B-Que Ribs won First Place at the National Bar-B-Que Convention Rib and Chicken Cook-off in Memphis, Tennessee. Or for another out-of-this-world dining experience, visit the Little A'Le'Inn (http://littlealeinn.com) in Rachel, Nevada, on your way to Area 51 (see Bucket List item 51). On your next visit to Albuquerque, New Mexico, have a meal or sweet treat at a Flying Star Café (http://flyingstarcafe.com). Wherever you live or travel, look around for spacey dining choices!

Flying Star Café sign in Albuquerque. Photo by Loretta Hall

62. See a rocket launch

No, not just on television. Be there to experience the actual event in sight, sound, and vibration. Go online for a schedule of upcoming launches at Cape Canaveral, Florida (https://www.kennedyspacecenter.com/events. aspx?type=rocket-launches); Vandenberg Air Force Base, California (http://spacearchive.info/vafbsked.htm); and Wallops Flight Facility, Virginia (https://www.nasa.gov/missions/ highlights/schedule.html#.VO0gHC7lyTY). Launch-viewing locations for each facility are described at https://www.nasa. gov/centers/kennedy/launchingrockets/viewing.html#. VO0h7i7lyTb.

A Delta II rocket launch at Cape Canaveral Air Force Station Launch Pad 17-B. Courtesy of NASA/Jerry Cannon. Robert Murray

61. Watch *Close Encounters of the Third Kind* and *Contact*

See descriptions and interesting trivia about these classic films at http://www.imdb.com/title/tt0075860 and http://www.imdb.com/title/tt0118884/?ref_=fn_al_tt_1, respectively.

Contact, released in 1997, is twenty years more recent than *Close Encounters*. Which film presents a more plausible scenario for communication with intelligent extraterrestrial beings? Which is simply more entertaining? Are those answers the same? You can also compare the *Contact* movie with the book of the same name (see Bucket List item 100). Carl Sagan and his wife wrote the story outline for the film. Were there substantial differences between the book and the movie? If so, which did you like better?

Devil's Tower. Wyoming. featured in Close Encounters of the Third Kind. *Public domain*

60. Hear the sounds of space

As they travel through our solar system, various space-craft detect radio waves and relay them to Earth, where they can be translated into sounds. Space probe Voyager 1 even captured the sound of crossing the boundary between our solar system and interstellar space. The sounds are as eerie and haunting as you might imagine. — https://www.youtube.com/watch?v=-MmWeZHsQzs, soundcloud.com/esaops and at www-pw.physics.uiowa.edu/space-audio.

Voyager space probe. Courtesy of NASA

59. Become a planetary cartographer

Study the surface of extraterrestrial objects in unprecedented detail at http://marstrek.jpl.nasa.gov. Mark and measure craters and linear surface features on Mercury photographed by NASA's Messenger spacecraft at http://cosmoquest.org/x/planet-mappers-tutorial. Identify and measure surface features on Mars at http://www.planetfour.org. Stay virtually closer to Earth and examine lunar features at http://www.moonzoo.org/how_to_take_part. Another option is https://cosmoquest.org (select "Moon Mappers" under "Do Science"); do the grand slam on this website and also examine Mars, Mercury, and the asteroid Vesta.

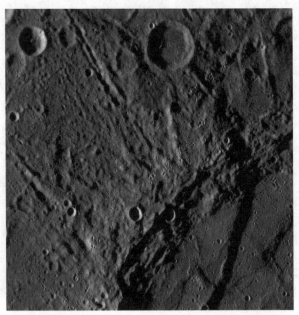

Mercury's surface photographed by Messenger. Courtesy of NASA

58. Try to get lucky

Search for contests that offer a chance to win a spaceflight ticket. Since 2005, companies including Volvo, 7UP, and KLM Royal Dutch Airlines have offered such prizes. For example, in 2009 the Guinness brewing company celebrated its 250th anniversary with a contest featuring a suborbital flight on Virgin Galactic as one of the prizes. In November 2014, people who bought a ticket to see the new movie *Interstellar* through the online Fandango site were entered in a contest to win a suborbital flight on XCOR Aerospace's Lynx Mark II vehicle. These contests come and go, so you will have to keep an eye out for them. A search for "space ticket contest" should do the trick.

Photo of a space shuttle reentry taken from the International Space Station. Courtesy of NASA

57. Download a space-themed screen saver or wallpaper

Explore space every time you return to your computer. For example, get a Space Travel screensaver at http://download.cnet.com/The-Space-Travel-Screensaver/3000-2257_4-10062298.html. Space.com offers hundreds of stunning extraterrestrial wallpapers (http://www.space.com/wallpapers). Download images taken with the Hubble space telescope at http://hubblesite.org/gallery/wallpaper.

Hubble space telescope wallpaper. Courtesy of Space Telescope Science Institute (STScI)

56. Read *Failure Is Not an Option*

In his book, NASA Mission Control Flight Director Gene Kranz gave a first-hand account of the Mercury, Gemini, and Apollo missions. In the atmosphere of the Cold War and the space race with the Soviet Union, some problematic events were kept out of the public discourse. Four decades later, Kranz tells it like it was. It is an appropriate reminder of the difficulty of space exploration as the new commercial space industry develops. — http://books.simonandschuster.com/Failure-Is-Not-an-Option/Gene-Kranz/9781439148815

This iconic slogan represented Mission Control's mindset during Apollo 13. Photo by Loretta Hall

55. Have a star named after you

Or after a loved one. Go to the International Star Registry and place your order. Proudly display your personalized certificate, which identifies the star and its galaxy along with the name you chose. You will also receive a colorful star chart with your star circled. — http:// starregistry.com

Or have your chosen star name and dedication message sent on a launch into space. Buy keepsake items to commemorate your star-naming adventure. — https://www. nameastarlive.com

Certificate commemorating the name of "Loretta" for the star officially designated Andromeda RA 0h 16m 21s D 27° 38'. Photo by Loretta Hall

54. Get stars in your eyes

Find a star no one has identified before. At the Stellar Classification Online Public Exploration (SCOPE) website, you can compare photographic images of star spectra and compare them to stars with known spectra. — http://scope.pari.edu

Participate in the Milky Way Project and catalog bubbles, star clusters, knots of gas, and dark nebulae in infrared images taken by the Spitzer Space Telescope. — http://www.milkywayproject.org

Join Galaxy Zoo (http://www.galaxyzoo.org) and help scientists understand galaxy formation by helping classify galaxies from images taken by the Sloan Digital Sky Survey. Search for undiscovered black holes at http://radio.galaxyzoo.org. A 2015 study of people participating in the black hole search found that "Trained volunteers are as good as professional astronomers at finding jets shooting from massive black holes and matching them to their host galaxies." You, too, can make astronomical history!

Infrared image from Spitzer Space Telescope. Courtesy of NASA/ESA/JPL-Caltech/UCLA/CXC/SAO

53. Wear space jewelry

You've heard of wearing your heart on your sleeve; here's a chance to wear your love of space as jewelry—cufflinks, necklaces, bracelets, earrings, dog tags, and more. Select from manmade items, including space organization logo items. — http://www.cafepress.com/+space-logo+jewelry

You can even wear a piece of outer space in jewelry displaying meteorites! — http://jurassicjewellery.co.uk, http://spacegems.org

Meteorite necklace. Courtesy of Jurassic Jewellery

52. See a space shuttle

Discovery, which flew the most missions, is on display at the Smithsonian National Air and Space Museum's Steven F. Udvar-Hazy Center in Chantilly, Virginia. *Atlantis*, the last shuttle to fly in space, is at the Kennedy Space Center Visitor Complex in Cape Canaveral, Florida. *Endeavour*, which was built after the shuttle *Challenger* disintegrated during launch, is at the California Science Center in Los Angeles. Enterprise, a prototype shuttle used for vibration tests and to fly approach and landing tests, is at the Intrepid Sea, Air & Space Museum in New York. See one, or see them all!

Find information about the missions of all space shuttles and specific display locations of the surviving ones at https://www.nasa.gov/mission_pages/shuttle/main/index.html.

Space shuttle Discovery. *Photo by Loretta Hall*

51. Visit Area 51

Well, you can't actually visit it, but you can get near it. It is in a remote desert location 83 miles north-northwest of Las Vegas, Nevada. Find out how to get near it and how to find the best legal viewpoint at http://visitarea51.com and at http://withoutbaggage.com/essays/area-51. Be sure to follow posted rules about taking photos. Those folks with the weapons are not kidding. Seriously.

Signs at the back gate to Area 51. Photo by Loretta Hall

50. Put yourself and the universe in perspective

Drive to a rural area on a clear night and look at the stars. Really look at them. As numerous as they are, the stars you can see are only a small part of the universe. They are millions or billions of miles away. In the time it has taken their light to reach Earth, they may have faded or exploded. New ones may have formed that you can't even see yet. What is it like out there now?

The constellation Sagittarius. Creative Commons License. photo by Scott Roy Atwood

49. Actively support space exploration

Put your money and your time where your dreams are. Join an organization that works to advocate for public and private space exploration. One example is the National Space Society (http://nss.org), an "independent, educational, grassroots, non-profit organization dedicated to the creation of a spacefaring civilization." Another is the Mars Society (http://www.marssociety.org), whose purpose is "to further the exploration and settlement of the Red Planet." Yet another, the Planetary Society (http://planetary.org), "sponsors projects that will seed innovative space technologies, nurture creative young minds, and be a vital advocate for our future in space."

National Space Society logo. Courtesy of National Space Society

48. Subscribe to free e-newsletters

Be informed about current events and discoveries in space. For example, read current and past issues of Kennedy Space Center's monthly magazine, *Spaceport News*. — https://www.nasa.gov/centers/kennedy/spaceport-magazine.html#.VPCvseHlyTY

And you can sign up for good newsletters at these websites:

http://spacenews.com
http://www.space.com
http://www.spacedaily.com
http://www.exploredeepspace.com
http://planetary.org/connect

A Daily Newsletter from SpaceNews

E-newsletter logo. Courtesy of SpaceNews.com

47. Display your frame of mind

Frame your favorite space photo or poster and hang it in your home or office. You can download high-resolution photos and have them professionally printed. For especially dramatic decor, download a six-panel mural up to 60 inches wide by 40 inches high. — http://hubblesite.org

Other high-resolution images suitable for framing are available at https://www.nasa.gov/multimedia/imagegallery/iotd.html#.VPCyB-HlyTZ.

Galaxy photographed by the Hubble space telescope. Courtesy of NASA, ESA, and the Hubble Heritage Team (STScI/AURA)

46. Wear logo apparel to spread awareness

Visibly and financially support your favorite space organizations by buying and wearing logo apparel. If you enjoy *The Space Show* (see Bucket List item 69), promote it by wearing a shirt or cap you purchase at Café Press. Virgin Galactic, the National Space Society, and XCOR also use Café Press to make and sell their logo wear. — http://www.cafepress.com

SpaceX merchandise is available at the company's own online store. — http://shop.spacex.com

Spaceport America merchandise is also sold on the organization's own website. — http://shop.spaceportamerica.com/Spaceport-merchandise.php

Other items, including neckties, are available at these websites: http://www.zazzle.com/space+logo+clothing?dp=252657725253948300; spaceshoponline.com/adultclothing.html; http://store.space.com

Logo shirts. Photo by Loretta Hall

45. Build a fantasy collection of space memorabilia

At Bonhams' annual space memorabilia auction, prices range from thousands of dollars to hundreds of thousands of dollars. However, you can get a catalog for the current year's "Space Memorabilia" auction for $40. — http://www.bonhams.com

RRAuction offers items costing a few hundred to a few thousand dollars. — http://www.preview.rrauction.com

Peruse their listings and select the items you would bid on if you had the financial resources (maybe you'll even come across something you can afford). Organize a group of fellow space buffs into a Fantasy Collectors League and bid against each other.

Mercury-era spacesuit. sold for $43,750. Courtesy of Bonhams

44. Visit the Goldstone Deep Space Communications Complex

Located in a remote location in California's Mojave Desert, it is one of three deep space tracking systems. Complementary sites in Spain and Australia enable uninterrupted tracking of spacecraft. One of the space probes the sites are tracking is New Horizons, which continues on beyond the solar system after its 2015 close encounter with Pluto. The two-and-a-half-hour Goldstone tours are free, but must be scheduled at least seventy-two hours in advance. — http://www.gdscc.nasa.gov/?page_id=35

Goldstone Deep Space Communications Complex, California. Courtesy of NASA

43. Remember or discover Space Food Sticks

Pillsbury developed these nutritious snacks for NASA's manned space program. Shaped like a long cigarette, their consistency was like stale Play-Doh, but they tasted better. They came in chocolate, chocolate mint, caramel, and peanut butter flavors. If you remember snacking on these in the 1970s, check off this item on your Bucket List.

If you don't remember them, or if you want to refresh your memory, watch the vintage television commercials. — https://www.youtube.com/results?search_query=space+food+sticks+commercial

Find out how they were developed for NASA. — http://www.spacefoodsticks.com/history.html.

After a brief revival in 2010, Space Food Sticks seem to have disappeared again. One can only hope they will return some day. In the meantime, you can make a facsimile for yourself using the recipe at www.cooks.com/recipe/pp7il8l8/space-food-sticks.html.

Vintage photo of Space Food Sticks package. Courtesy of General Mills

42. Eat Astronaut Ice Cream

Unlike Space Food Sticks, Astronaut Ice Cream has endured. It is available at the gift shops of most museums that offer space exhibits. With a consistency of Styrofoam, it does taste like ice cream, but without the coldness and moisture. Definitely a different taste sensation! — http://www.think-geek.com/product/9e07

If you're thirsty after eating the dry ice cream, slurp some Tang instant orange-flavored drink also used by astronauts. — http://heliopoli.com/2007/12/02/tang-a-correlative-history

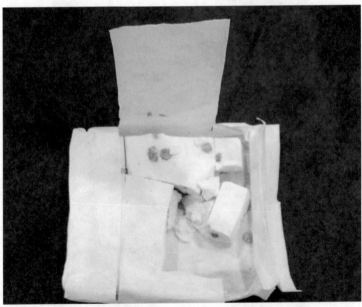

Astronaut Ice Cream (mint chocolate chip). Photo by Loretta Hall

41. Carry the astronaut food theme to fruition

Prepare an entire on-orbit meal. You will find all the information you need in *The Astronaut's Cookbook: Tales, Recipes, and More* by Charles Bourland and Gregory Vogt. It includes a history of space foods, explanations of requirements for space foods, and recipes for cooking your own. Don't worry, you won't have to suit up to eat your meal. — http://www.springer.com/us/book/9781441906236

Learning to eat like an astronaut. Photo by Loretta Hall

40. Be a model space buff

Build a collection of scale models of historic and modern spacecraft. — http://www.revell.com, http://www.realspacemodels.com.

Experienced model builders might want higher-end versions. — http://www.tailwinds.com/space-and-nasa-scale-models.html.

For fun with kids, download free kits, print them on card stock, and build them together. — http://www.jpl.nasa.gov/scalemodels, http://spacecraftkits.com/free.html

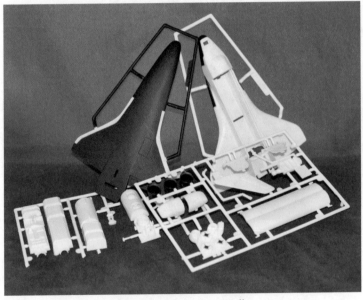

Model kit for a space shuttle. Photo by Loretta Hall

39. Explore our solar system from the comfort of your home

NOVA offers an online, interactive tour of our solar system. Watch the planets move around the sun in their respective orbits. Select a planet (or our Moon) and examine its entire surface. See the relative positions of the planets at any time you choose. Move planets at will! But don't let that power go to your head, because the laws of orbital mechanics still prevail. — http://www.pbs.org/wgbh/nova/space/tour-solar-system.html

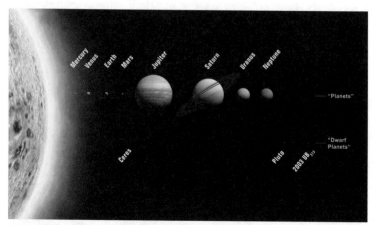

Our solar system. Courtesy of NASA

38. Host an event during World Space Week

Established in 1999 by the United Nations General Assembly, this celebration is held internationally between October 4 and October 10 every year. Find out the current year's theme and start planning. That website also suggests event ideas and offers instructions for organizing an official event. It can be as simple as a star party, a costume party, a classroom project, or a public lecture. — http://www.worldspaceweek.org

Logo. Courtesy of World Space Week Association

37. Visit Johnson Space Center

Houston, you will have no problem visiting Johnson Space Center. The space shuttle you see as you approach the building is a full-scale replica built after the shuttle program ended. Inside the building, you will find exhibits, interactive presentations, live demonstrations, and historic and modern Mission Control Centers. For a special treat, sign up for the Level Nine Tour. Only twelve people per day can take this five-hour, behind-the-scenes tour of Johnson Space Center. Advance reservations are required, and participants must be at least fourteen years old. — http://spacecenter.org

Mission Control Center. Courtesy of Johnson Space Center

36. Play a space adventure game

You can even play vintage MS-DOS games free online. Titles include "Star Wars" and "Star Trek." — https://archive. org/details/softwarelibrary_msdos_games

Download NASA's free "Moonbase Alpha" game. — https://www.nasa.gov/offices/education/programs/national/ ltp/games/moonbasealpha/index.html

Download "Orbiter," a 1986 space shuttle flight simulator based on NASA procedures. — http://old-games.com/download/4362/orbiter

Or download a newer, free 2000 "Orbiter" game that claims it "offers accurate physics, excellent 3D graphics, astronomy features, and a first-person astronaut's perspective."— http://orbit.medphys.ucl.ac.uk

If you prefer a multi-player strategy game, buy "Space Program Manager: Road to the Moon," which was developed in cooperation with Apollo 11 astronaut Buzz Aldrin. — http:// matrixgames.com/products/462/details/Buzz.Aldrins.Space. Program.Manager

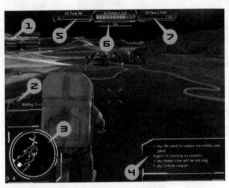

Moonbase Alpha screen shot. Courtesy of NASA

35. Have lunch with an astronaut

Spend an hour chatting with a veteran astronaut over lunch. This experience is offered at both Johnson Space Center in Houston and Kennedy Space Center in Cape Canaveral. Tickets must be purchased in advance. Check a schedule of which astronauts will be available for lunch at the centers' websites. — http://spacecenter.org/attractions/lunch-with-an-astronaut, https://www.kennedyspacecenter.com/the-experience/lunch-with-an-astronaut.aspx

Lunch with an astronaut. Courtesy of NASA. JSC

34. Attend space camp

Space camps for kids are available at many locations around the country. The US Space and Rocket Center in Huntsville, Alabama, also offers Adult Space Academy®. Three- and four-day programs cost $500 to $600, including meals and lodging. Along with an education in the history of spaceflight, you get to train on astronaut simulators and build a model rocket. — http://spacecamp.com/space/adult

Or for a different kind of experience, try the Astronomy Camps for Adults at the University of Arizona's Steward Observatory in Tucson. They offer Beginning and Advanced options lasting three or four days and costing $600 to $700. — http://astronomycamp.org/pages/adultcamp.html

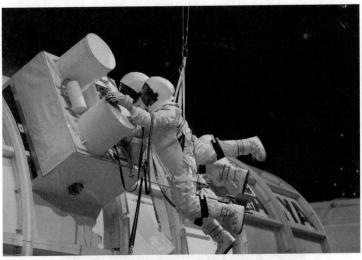

Adult Space Camp®. Courtesy of Space Camp®

33. Simulate a trip to the moon, timewise

The 1970 Apollo 13 mission really dramatized the length of a round trip to the Moon, as the world watched and waited to see whether the three astronauts in the crippled spacecraft would return safely to Earth. Consciously experience that length of time by noting a starting point and be aware of all you do for the next 5 days, 22 hours, 54 minutes, and 41 seconds. For example, if you start on a Monday morning at 8 o'clock, you will finish the flight-duration period the following Sunday morning, about 5 minutes before 7 o'clock. You can check your progress with NASA's Apollo 13 timeline. Think about spending that entire time in a confined space, using minimal power for lighting and heat, wondering whether you will survive. — http://history.nasa.gov/SP-4029/Apollo_13h_Timeline.htm

Earth as seen from Apollo 8. Courtesy of NASA

32. Mentally prepare for a space shuttle mission

Study NASA's 1988 *Shuttle Reference Manual*. It gives a detailed mission profile, explains abort procedures for various scenarios, describes the components of the spacecraft and launch vehicles, lists the equipment available for the crew, and explains the systems for on-orbit flight. Finally, a Mission Events Summary guides you through a complete flight. — http://spaceflight.nasa.gov/shuttle/reference/shutref/verboseindex.html

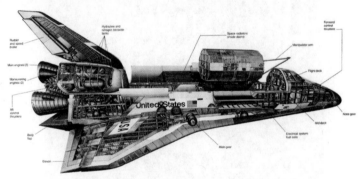

Space shuttle diagram. Courtesy of NASA

31. Read *Packing for Mars*

Mary Roach's 2010 book, subtitled *The Curious Science of Life in the Void*, is a fascinating, often humorous look at life in space and space simulations on Earth. You will learn about things you never dreamed of. Still undecided about the question posed in Bucket List item 93, about the Mercury 13 "expediency"? There's even a brief explanation of that in *Packing for Mars*, along with other nitty-gritty details. — http://maryroach.net/packing-for-mars.html

Learning to pack light and tight. Photo by Loretta Hall

30. Visit the International UFO Museum and Research Center

Roswell is well known as the location of a purported flying saucer crash in 1947. The UFO museum in the southeastern New Mexico town offers displays depicting reports of the Roswell Incident as well as other unidentified flying object events. A well-equipped library is available for researchers, students, and authors seeking information about worldwide UFO phenomena. — http://roswellufomuseum. com

International UFO Museum and Research Center. Photo by Loretta Hall

29. Visit the other Roswell space museum

Farther up Main Street, you will find the Roswell Museum and Art Center. A large windmill frame outside the museum is actually the launch tower Robert Goddard used for his liquid-fuel rocket research near Roswell during the 1930s. Inside, you will find a wing of the museum devoted to Goddard's work. A highlight of the exhibits is a replica of Goddard's workshop, filled with equipment he and his assistants used. — http://roswellmuseum.org/current/dr-robert-hutchings-goddard-collection

Robert Goddard's workshop recreation. Photo by Jerry Hall

28. Visit the New Mexico Museum of Space History

After visiting Roswell, take a road trip tour of other New Mexico sites that played important roles in the development of manned spaceflight. Begin with a stop in Alamogordo (117 miles southwest of Roswell) at the New Mexico Museum of Space History Although the museum highlights New Mexico contributions (including training chimpanzees for NASA's Mercury program), it is a comprehensive overview of space history. One display features inductees in the museum's International Space Hall of Fame. — http://www.nmspace-museum.org

An apple on the gravestone of Ham. the astrochimp at the New Mexico Museum of Space History. Photo by Loretta Hall

27. Visit the White Sands Missile Range Museum

From Alamogordo, drive 42 miles southwest across the Tularosa Basin and its famous White Sands National Monument. Just before you cross the San Augustin Mountains to reach Las Cruces, stop at the White Sands Missile Range Museum. More than fifty objects are displayed in the outdoor missile park. Inside the museum are displays about the history of the region, including rocket research done at White Sands Missile Range. In a separate building, you will see one of eight complete V-2 missiles on display in the United States. Some sections of the shell have been removed so you can see internal components. — http://wsmr-history.org

White Sands Missile Range rocket display. Photo by Loretta Hall

26. Visit Spaceport America

The strangely named town of Truth or Consequences, New Mexico, 75 miles north of Las Cruces, is the starting point for Spaceport America tours. The world's first purpose-built commercial spaceport is unique with its isolated, historic setting, high elevation (4,595 feet), and its futuristic architecture. Unmanned, suborbital research rockets are launched a few times a year from Spaceport America. Space tourism flights are expected to begin in 2017 or so. — http://spaceportamerica.com/experience

Spaceport America terminal and hangar facility (the Virgin Galactic Gateway to Space). Photo by Loretta Hall

25. Visit the Very Large Array

After your tour of Spaceport America, drive north 75 miles to Socorro, New Mexico, then 50 miles west to see the Karl G. Jansky Very Large Array. A moveable array of 27 huge radio telescope antennas spanning an area up to 23 miles across creates an other-worldly environment. You may recognize it from watching *Contact* (see Bucket List item 61). — http://public.nrao.edu/tours/visitvla

Less spectacular but also interesting is a closely spaced grid of skeletal antennas that comprise one station of the Long Wavelength Array on the same property. — http://phys.unm.edu/~lwa/abouthome.html

Part of the Very Large Array. Photo by Loretta Hall

24. Explore an Earthly Moonscape

In the 1960s and early '70s, Apollo astronauts took geology field trips to many places with features similar to what they might find on the Moon. For example, Neil Armstrong was one of nine astronauts who explored the San Francisco Volcanic Field near Flagstaff, Arizona, in 1963. Other astronauts trained at the Grand Canyon in Arizona. The Rio Grande Gorge near Taos, New Mexico, was the closest analogy on Earth to Hadley Rille, a lunar destination of the Apollo 15 crew. — http://www.hq.nasa.gov/alsj/of2005-1190_table1.pdf

Rio Grande Gorge astronaut training site. Photo by Loretta Hall

23. Read old magazine articles about astronauts

These articles place the Mercury, Gemini, and Apollo events into a social context you won't get from a history book. For example, a 1960 *Time* magazine article about a female pilot (see Bucket List item 93) who successfully underwent the same physical examination as the male Mercury astronaut candidates described her as a "slender (5 ft. 7 in., 121 lbs.) blonde" "bachelor girl" and gave her vital statistics: 36-27-34. Besides the articles about the astronauts and their families, read other articles and advertisements that reflect that era's culture. Find magazines in libraries, or scour antique stores and collect them.

Vintage Life *magazines. Photo by Loretta Hall*

22. Hand it to the astronauts

The Moon explorers wore gloves that Apollo 17 astronaut Harrison Schmitt described as being "like a balloon, the shape of your hand, and 3.8 pounds per square inch." They were bulky, and required finger and forearm strength. To get some sense of that, put on a pair of welding gloves and try striking a rock with a hammer. Put on a bulky coat to simulate the rest of the space suit, and use your gloved hands to push a small pole into the ground. Use a long-handled scoop to pick up a bit of soil and pour it into a plastic bag. Climb a 6-foot ladder. Grip a steering wheel.

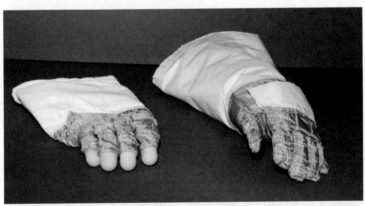

Extravehicular Activity (EVA) gloves. Courtesy of NASA

21. Visit Kennedy Space Center

Cape Canaveral, Florida, has been the focal point of America's manned space launches since the Mercury era. At Kennedy Space Center, you will find great exhibits and activities. Opt for the extra Shuttle Launch Experience, which is described as "an all-too-real simulation of the space shuttle's eight-and-a-half-minute ascent into orbit." Then there's a momentary feeling of weightlessness, followed by a stunning view of the Earth as the shuttle's cargo bay doors open. — http://kennedyspacecenter.com/the-experience/shuttle-launch-experience.aspx

Shuttle Launch Experience. Courtesy of NASA, KSC

20. Train as an astronaut

Come back for a second day at Kennedy Space Center for a half-day-long Astronaut Training Experience. Practice a rendezvous and docking with the International Space Station. Get shuttle flight and landing training, experience a simulation of Moon gravity, and a ride on the Multi-Axis Trainer. About as close as you can get to the real thing. — http://kennedyspacecenter.com/the-experience/astronaut-training-experience.aspx

Astronaut Training Experience. Courtesy of NASA, KSC

19. Virtually explore the Moon

Install Google Earth on your computer. With the application running, click the icon of a ringed planet from the toolbar at the top of the screen and select "Moon." You will be able to explore the Moon's entire surface and zoom in on sites for a closer look. Locations of known manned and unmanned landing/impact sites are marked. Bonus: you can also use Google Earth to explore Mars in detail. — http://www.google.com/earth

However, the locations of a number of hardware crash sites on the Moon are not known yet. See if you can be the first person to find the resting place of one of a dozen artifacts including the Apollo 10 Lunar Module descent stage or the Apollo 11 Lunar Module ascent stage. — http://www.space.com/25613-moon-far-side-nasa-spacecraft-crash.html

The Moon' surface. Courtesy of NASA/USGS/JAXA/SELENE via Google Earth Pro

18. Take two looks at *Cosmos*

Watch all of Carl Sagan's 1978-79 television series *Cosmos: A Personal Voyage* and all of Neil deGrasse Tyson's 2014 series *Cosmos: A Spacetime Odyssey*. Both are available for purchase, and you can watch the Sagan series free on YouTube. Compare and contrast. How did our understanding of the universe change during those thirty-five years? Which was more effective at conveying the excitement and importance of space research to the general public? — https://www.youtube.com/playlist?list=PLBA8DC67D52968201, http://www.haydenplanetarium.org/tyson/buy/videos/cosmos-a-spacetime-odyssey

Cosmos: A Personal Voyage *logo. Courtesy of PBS*

17. Envision a scale model of our solar system

But be careful what scale you choose. If the Earth were the size of a baseball (2.9 inches), the Moon would be 0.8 inches in diameter and 7.3 feet away from the Earth. Mars would be the size of a ping pong ball (1.5 inches) and a quarter of a mile from Earth. The Sun would be 26.4 feet in diameter and half a mile from the Earth. Get corresponding values for the rest of the solar system at. — http://thinkzone.wlonk.com/SS/SolarSystemModel.php

NASA even provides instructions for building a realistic scale model. — http://nasa.gov/pdf/546148main_ESS8_ScaleModelsOfTheSolarSystem_C3_Final.pdf

Our solar system. Courtesy of NASA

16. Study the Sun

At NOVA Labs, you can learn about solar activities and then use images from some of NASA's best solar space telescopes to examine events on the sun. See if you can predict a solar storm. Share your observations and theories with other members of the Sun Lab. — http://pbs.org/wgbh/nova/labs/lab/sun

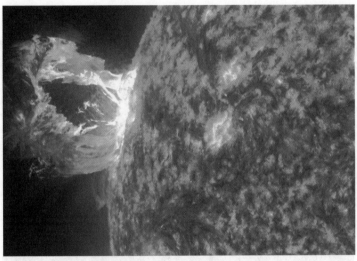

A solar storm. Courtesy of NASA

15. Get a handle on space junk

Learn about orbital debris. How much of it is there? Where does it come from? What hazard does it create for spacecraft? Can any of it survive re-entry through the atmosphere? — http://orbitaldebris.jsc.nasa.gov/index.html

See what objects will be the next to fall from orbit. — http://www.satview.org/spacejunk.php

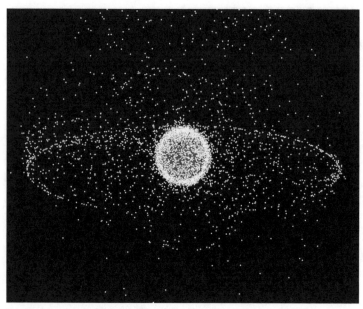

Computer image of orbital debris (not to scale). Courtesy of NASA

14. Take a Zero-G airplane flight

Weightlessness—more precisely, microgravity—is one of the most difficult space conditions to simulate on Earth. On a Zero-G airplane flight, a modified Boeing 727 airplane flies a succession of vertical parabolic paths. In the unobstructed cargo bay, you will experience fifteen periods of weightlessness, each lasting about half a minute. Two other flight paths allow you to experience the gravitational force on the Moon (one-sixth of Earth's gravity) and on Mars (one-third of Earth's gravity). Flights are scheduled in major cities including New York, Chicago, Boston, Miami, Las Vegas (Nevada), and Seattle. See the website for locations and dates. — http://gozerog.com

Activity in microgravity. Courtesy of Zero-G

13. Look for interstellar particles in aerogel

NASA's Stardust spacecraft was launched in 1999 and its sample capsule returned to Earth in 2006. In the midst of its flight, it exposed its collector to streams of interstellar dust. In the Stardust@home project, scientists invite volunteer observers to examine images of the collector and search for interstellar particle traces. Here's an idea of the enormity of the task: "They are tiny, only about a micron (a millionth of a meter) in size! These minuscule particles are embedded in an aerogel collector 1,000 square centimeters in size. To make things worse, the collector plates are interspersed with flaws, cracks, and an uneven surface. All this makes the interstellar dust particles extremely difficult to locate." Give it your best shot! — http://stardustathome.ssl.berkeley.edu

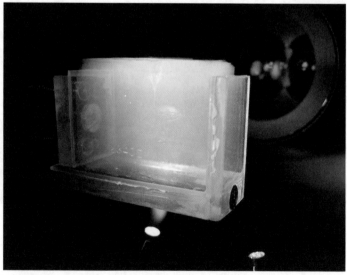

Aerogel from Stardust spacecraft. Courtesy of NASA

12. Find out if you are smarter than a chimpanzee

When Enos flew the first occupied orbital flight of the Mercury program, he was presented with four types of problems on a rotating schedule. He would work on each task for 12 minutes and then rest for 6 minutes before being presented with the next problem set. The following were the four types of problems:

Once every 2 minutes or so, a blue light came on, and he had to pull the left-hand lever within 5 seconds. Also, when a white light came on, he had to pull the right-hand lever within 15 seconds.

When a green light came on, he had to wait exactly 20 seconds and then pull an associated lever.

He had to pull a lever exactly fifty times.

He had to identify the odd symbol in a group of three that appeared on a screen.

Get a friend to help you simulate those tasks and time your responses. Can you do each task correctly for even 5 minutes? — http://history.nasa.gov/SP39Chimpanzee.pdf

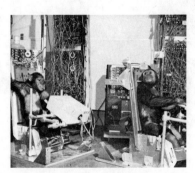

Ham (left) and Enos training for Mercury suborbital and orbital flight activities. respectively. Courtesy of U.S. Air Force

11. Watch *Apollo 13*

The 1995 movie, starring Tom Hanks and directed by Ron Howard, was widely acclaimed as being an accurate depiction of the 1970 mission that was crippled by an explosion on its way to the Moon. In the film, actor Ed Harris portrayed NASA flight director Gene Kranz. Kranz said, "The one thing that never happened is that Ed Harris sort of loses his temper when the thermal re-entry procedures don't come in. He yells at some of the people in the room, and he kicks a waste basket. That never happened, because you never lose control of the situation." Other than that small but significant error, the movie was well researched. — http://www.imdb.com/title/tt0112384

Apollo 13 insignia. Courtesy of NASA

10. Experience a flotation tank

The neutral buoyancy of a flotation tank mimics the feeling of weightlessness. The isolation inside a flotation pod simulates confinement in a small space capsule. Drift effortlessly, imagining yourself on the way toward Mars.

Learn about flotation therapy. — http://serenedreams.com/

Find a location near you where you can try it. — http://www.floatation.com/wheretofloat.html

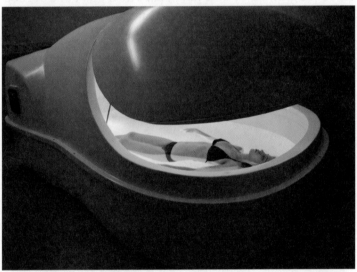

Float pod. Courtesy of Floatpod.com

9. Smell space

Space is not the empty vacuum people used to believe it was. Space dust and ionized particles flow through it. As a result of being exposed to that environment, astronauts re-entering their spacecraft experience an unusual aroma. They report that it is difficult to describe, but they liken it to hot metal, arc welding fumes, or spent gunpowder. Others say it reminds them of seared steak or dampened charcoal. That combination is easy to reproduce. Fire up a charcoal barbeque grill, throw on a juicy steak, then pull out a burned briquette and spray it lightly with water. Mmmm!

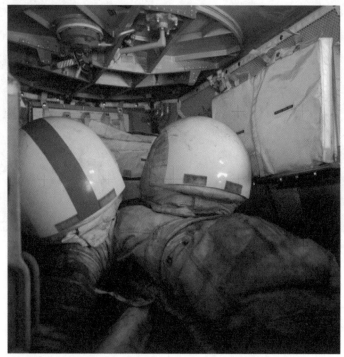

Moon dust on Apollo 17 EVA suits. Courtesy of NASA

8. Find a previously unknown planet

Search for undiscovered planets by examining how the brightness of a star changes over time, using about thirty days of observations from the Kepler spacecraft. — http://www.planethunters.org.

Known exoplanets (those outside our solar system) are given official names like "PSR 1257+12-b," but you can help give some of them unofficial names that are more appealing! — http://www.uwingu.com

To further scientists' understanding of how planets form, hunt for dusty debris disks around stars that are in the early stages of planet formation. — http://www.diskdetective.org

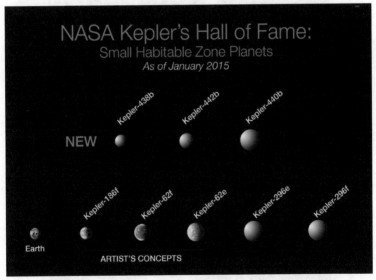

Small habitable-zone exoplanets discovered by the Kepler space telescope. Courtesy of NASA

7. Discover the vision of Wernher von Braun

Wernher von Braun and others wrote a series of articles in eight issues of *Colliers* magazine between March 22, 1952, and April 30, 1954. They wanted to convince the public and, indirectly, government leaders that manned exploration of space was achievable in the near future. Read the articles, which are archived online. Dates of the issues containing the space articles are 1952 (March 22, October 18, and October 25), 1953 (February 28, March 7, March 14, June 27), and 1954 (April 30). — http://unz.org/Pub/Colliers

Dr. Wernher von Braun. director of Marshall Space Flight Center. 1960-1970. Courtesy of NASA/MSFC

6. Wear a real space suit

The space suit design firm Final Frontier Design offers a two-hour experience including wearing an actual space suit, a simulated launch and flight in the pressurized suit, and testing prototype gloves in a negative pressure glove box. See if you can swing a golf club in a space suit like Alan Shepard did on the Moon during the Apollo 14 mission. The experience, at the company's design studio in Brooklyn, costs about $500. — http://www.finalfrontierdesign.com/sse

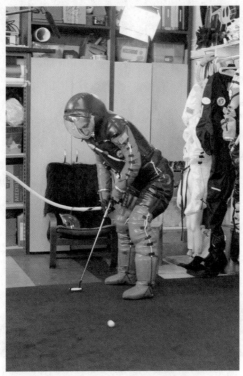

Hubble space telescope repairman. astronaut Mike Massimino. golfing in a space suit. Courtesy of Final Frontier Design. ©2014

5. Get training in space suits and systems

The National AeroSpace Training and Research (NAS-TAR) Center in Pennsylvania offers a one-day course that includes topics such as altitude physiology, communications, mobility/maneuverability, and emergency scenarios. The day culminates with a "flight" in an altitude chamber. — http://www.nastarcenter.com

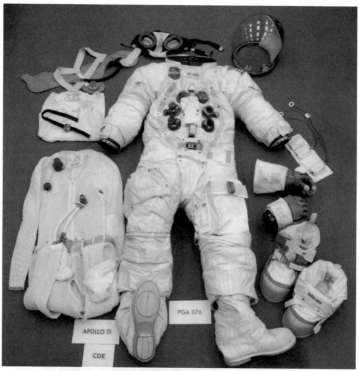

Neil Armstrong's extravehicular mobility unit (EMU) space suit from Apollo 11. Courtesy of NASA

4. Help search for extraterrestrial intelligence

Volunteer your computer's spare time to help analyze signals from radio telescopes for patterns that may represent a transmission from an extraterrestrial civilization. Download software that detects idle time on your computer, imports data, analyzes it, and reports results. The SETI@ home (Search for Extraterrestrial Intelligence) project operates from the University of California at Berkeley. — http:// setiathome.berkeley.edu

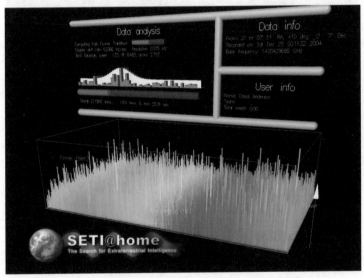

Screen shot of SETI@home data graphics. Courtesy of SETI@home

3. Experience human factors training for commercial spaceflight

Sirius Astronaut Training offers training in sensorimotor human factors such as motion sickness, spatial disorientation and movement errors, spatial illusions, and artificial gravity/ rotating environment Sign up for a day or two (Protocol A) or a week or two (Protocol B) at the Waltham, Massachusetts, facility. — http://siriusastronauttraining.com

Brandon Hogan rides a spinning Barany chair designed to teach astronauts about spatial disorientation. Courtesy of NASA/David C. Bowman

2. Arrange for a celestial burial

Before you pass from your physical life, let your loved ones know you want some of your cremains to be flown into space. Or arrange it yourself through a preplanned agreement. Options include a suborbital flight with the cremains returned to relatives after the flight, an Earth-orbit flight, launch to lunar orbit, launch to lunar impact, or an unending journey into deep space. On the Earth-orbit flights, the capsule containing cremains may burn up on eventual re-entry into the atmosphere, or the cremains can be dispersed in orbit and re-enter the atmosphere as individual particles. Prices start around $1,300. — http://elysiumspace.com, http://celestis.com

Courtesy of Elysium Space

1. Take a trip to space while you're alive to enjoy it

A ticket for a two-and-a-half hour suborbital flight on Virgin Galactic's SpaceShipTwo costs $250,000 and allows you to cavort in weightlessness (microgravity) for about four minutes. — http://virgingalactic.com

If you want an Apollo-style launch experience, choose a ride in Blue Origin's New Shepard; after separating from the reusable booster rocket, the capsule coasts for four minutes in weightlessness before parachuting back to Earth. The ticket price was not available at press time. — https://www.blueorigin.com

A strapped-in, copilot-seat ride in an XCOR Lynx II runs about $150,000 for a thirty-minute suborbital flight. — http://xcor.com

For a more leisurely voyage, take a four-to-five-hour flight to the edge of space (above 99 percent of the Earth's atmosphere) in a World View balloon for $75,000. Look into the mid-day blackness of space, marvel at the thin, blue layer of atmosphere surrounding the Earth, and ponder the immensity of the universe. http://worldviewexperience.com

Note: All ticket prices are subject to change.

Test flight of Virgin Galactic's SpaceShipTwo. Courtesy of Virgin Galactic

Some Last Words

Wasn't that an amazing exercise, just reading through this *Space Buff's Bucket List*? How many did you realize you've already done? So far, I've done about half of them. While writing this book, I discovered possibilities I'd never thought of before, and I'm looking forward to checking off more items for myself.

I hope you've discovered some new possibilities, too, along with the enthusiasm for actually doing them. Go ahead and write in this book. Check off things you have already done. Mark other things you plan to do. Use a different mark for things you'd like to do if you ever got the opportunity (like being able to afford a ticket to space).

Even though the title of this book is *The Complete Space Buff's Bucket List*, I know in my heart that it isn't really complete. In fact, a couple of friends of mine have already suggested some things I hadn't thought to include, and I've even thought of a few more—see "Near Misses" on the next page. Following that section, there are lists of more books and magazines to read.

Then it's your turn to get creative. Think up your own space buff's bucket list and write the entries on the pages provided below. If you come up with some really cool ideas, email them to me (Loretta@AuthorHall.com).

I hope this is the start of a beautiful experience!

Near Misses

Watch the International Space Station fly over. — *Ted Spitzmiller*

Have a meal in the Space Needle in Seattle, Washington. — *Jerry Hall*

Locate the Tranquility Base site on the Moon through a telescope. — *Ted Spitzmiller*

If you're a sports fan, root for a space-named team like the Houston Rockets or the Phoenix Suns in pro basketball, the major-league Houston Astros or the minor-league Huntsville Stars in baseball, or the Los Angeles Galaxy in major-league soccer. — *Jerry Hall*

Buy land on the Moon (well, at least a certificate naming you as landowner) at http://www.lunarland.com or at http://lunarregistry.com. MoonEstates also offers land on Mars and Venus at http://moonestates.com. Or go grand and purchase a full square mile on Uranus for $19.95 plus shipping (for the certificate) at http://buyuranus.com — *Loretta Hall*

Watch the National Geographic Channel's *StarTalk* on television or online — *Barbe Awalt*

Buy three-dimensional sculptures of astronomical objects embedded in clear cubes of crystal glass at http://crystalnebulae.co.uk. — *Loretta Hall*

Watch a meteor shower. Find out when and where they will be visible at https://stardate.org/nightsky/meteors. — *Loretta Hall*

Explore the intricacies of domestic and international space law at http://www.spacepolicyonline.com/space-law. — *Loretta Hall*

Space Books to Read

Any astronaut's memoir — for example:

First on the Moon: A Voyage with Neil Armstrong, Michael Collins and Edwin E. Aldrin, Jr. (Little, Brown, 1970)

John Glenn: A Memoir (Bantam, 2000)

Animals in Space: From Research Rockets to the Space Shuttle by Colin Burgess and Chris Dubbs (Springer-Praxis, 2007)

Apollo Moon Missions: The Unsung Heroes by Billy Watkins (Bison Books, 2007)

Contact by Carl Sagan (Pocket Books, 1997, ©1985)

Cosmos by Carl Sagan (Ballantine Books, 2013, ©1980)

Countdown: A History of Space Flight by T.A. Heppenheimer (Wiley, 1999)

Death by Black Hole: And Other Cosmic Quandaries by Neil deGrasse Tyson (W.W. Norton, 2014)

Dr. Space: The Life of Wernher von Braun by Bob Ward (Naval Institute Press, 2005)

Earthrise: How Man First Saw the Earth by Robert Poole (Yale University Press, 2008)

Failure Is Not an Option by Gene Kranz (Simon & Schuster, 2000)

Homesteading Space: The Skylab Story by David Hitt, et al. (Bison Books, 2011)

The Martian by Andy Weir (Broadway Books, 2014)

Mission to Mars: My Vision for Space Exploration by Buzz Aldrin (National Geographic, 2013

Moonshots & Snapshots of Project Apollo by John Bisney and J.L. Pickering (University of New Mexico Press, 2015)

NASA's Scientist-Astronauts by David J. Shayler and Colin Burgess (Springer-Praxis, 2007)

Origins: Fourteen Billion Years of Cosmic Evolution by Neil deGrasse Tyson and Donald Goldsmith (W.W. Norton, 2014)

Out of this World: New Mexico's Contributions to Space Travel by Loretta Hall (Rio Grande Books, 2011)

Packing for Mars: The Curious Science of Life in the Void by Mary Roach (W.W. Norton, 2010)

The Pre-Astronauts: Manned Ballooning on the Threshold of Space by Craig Ryan (Naval Institute Press, 2003)

The Right Stuff by Tom Wolfe (Picador, 2008)

Right Stuff, Wrong Sex: America's First Women in Space Program by Margaret A. Weitekamp (Johns Hopkins University Press, 2005)

Rocket Man: Robert H. Goddard and the Birth of the Space Age by David A. Clary (Hyperion, 2003)

Selecting the Mercury Seven: The Search for America's First Astronauts by Colin Burgess (Springer-Praxis, 2011)

Space and the American Imagination by Howard E. McCurdy (Johns Hopkins University Press, 2011)

Space Chronicles: Facing the Ultimate Frontier by Neil deGrasse Tyson (W.W. Norton, 2013)

Space Pioneers: In Their Own Words by Loretta Hall (Rio Grande Books, 2014)

Spaceshots & Snapshots of Projects Mercury & Gemini by John Bisney and J.L. Pickering (University of New Mexico Press, 2015)

This New Ocean: The Story of the First Space Age by William
    E. Burrows (Modern Library, 1999)

Space Magazines

Ad Astra (membership magazine of the National Space Soci-
    ety, nss.org)

Air and Space/Smithsonian (airspacemag.com)

Astronomy (astronomy.com)

KSC's Spaceport (nasa.gov/centers/kennedy/spaceport-mag-
    azine.html#.VPCvseHlyTY)

Quest: The History of Spaceflight Quarterly (spacehistory101.
    com)

RocketSTEM (rocketstem.org)

Sky and Telescope (skyandtelescope.com)

Not-To-Miss Space Movies

2001: A Space Odyssey

2010: The Year We Make Contact

Apollo 13

Close Encounters of the Third Kind

Contact

Gravity

The Martian

Here's What They Said….

As you think about the excitement and value of space travel, consider these remarks from former astronauts about their experiences:

Launch

"While you're awaiting launch, you're going to be in the grips of two fundamental emotions, one of which is gut fear. You will fear for your life while you're out there; it's a dangerous business, and you know things can go really, really bad. But at the same time you're gripped with this intense fear, you're simultaneously overwhelmed with boundless joy." — Mike Mullane, space shuttle mission specialist, in an October 2015 speech

"The rocket was huffing as puffs of vapor vented from it; the tanks were continually topped off. The Saturn V reminded me of a tethered animal pawing at the ground, ready to run. It no longer seemed like a large chunk of metal—it appeared to fume with frustration, ready to be unleashed, unrestrained." — Al Worden, Apollo 15, from *Falling to Earth*

Mike Mullane aboard a space shuttle. Courtesy of NASA

On the Moon

"I was surprised by the apparent closeness of the horizon. I was surprised that … there was no dust when you kicked. You never had a cloud of dust there. That's a product of having an atmosphere, and when you don't have an atmosphere, you don't have any clouds of dust." — Neil Armstrong, Apollo 11, in a September 2001 oral history interview at NASA Johnson Space Center

"The gloves are like a balloon the shape of your hand and 3.8 pounds per square inch. So any time you want to pick something up, you have to work against that pressure. That works the forearm muscles. We didn't really understand that we should have had specific, heavy-duty, pre-mission training for a long period of time to get those muscles in shape." — Harrison (Jack) Schmitt, Apollo 17, in a July 2012 speech

A lunar footstep. Courtesy of NASA

Future

"I think space travel has a heck of a future. I just hope we will participate. And I think we will…. Deep space travel is very different than orbital travel. Orbital travel has become almost routine; deep space is probably never going to become routine." — Harrison (Jack) Schmitt, Apollo 17, in an October 2014 speech

"I am absolutely convinced that there's a genetic drive in humans that says, 'We have got to develop a capability to go somewhere else when we can't live here anymore.'" — Al Worden, Apollo 15, in an October 2015 speech

Jack Schmitt and Loretta Hall, July 2012. Courtesy of Loretta Hall

Build Your Own Bucket List

1.
2.
3.
4.
5.
6.
7.
8.
9.
10.
11.
12.
13.
14.
15.
16.
17.
18.
19.
20.
21.
22.
23.
24.
25.
26.
27.
28.
29.
30.
31.
32.

33.
34.
35.
36.
37.
38.
39.
40.
41.
42.
43.
44.
45.
46.
47.
48.
49.
50.
51.
52.
53.
54.
55.
56.
57.
58.
59.
60.
61.
62.
63.
64.
65.
66.

67.

68.

69.

70.

71.

72.

73.

74.

75.

76.

77.

78.

79.

80.

81.

82.

83.

84.

85.

86.

87.

88.

89.

90.

91.

92.

93.

94.

95.

96.

97.

98.

99.

100.

The National Space Society

Each book in the Bucket List Series shines a spotlight on a relevant nonprofit organization. In this case, it is the National Space Society (NSS).

NSS is an independent, educational, grassroots, non-profit organization dedicated to the creation of a spacefaring civilization. It emerged in 1987 when the National Space Institute, which was founded in 1974, merged with the L5 Society, which was founded in 1975. With more than fifty chapters in the United States and in other countries, NSS is well respected as a citizen's voice on space endeavors.

The NSS website states the organization's mission this way: "The Mission of NSS is to promote social, economic, technological, and political change in order to expand civilization beyond Earth, to settle space and to use the resulting resources to build a hopeful and prosperous future for humanity. Accordingly, we support steps toward this goal, including human spaceflight, commercial space development, space exploration, space applications, space resource utilization, robotic precursors, defense against asteroids, relevant science, and space settlement oriented education."

Under the US tax code, NSS is a tax-exempt, educational organization with a 501(c)(3) designation. Visit its website for information about joining, donating, and/or purchasing logo items. — http://nss.org

ABOUT THE AUTHOR

Loretta Hall is a freelance writer and author of nonfiction books. She has contributed hundreds of articles to a variety of publications including *USA Today Special Publications*, *RocketSTEM* magazine, and *Discoveries in Modern Science: Exploration, Invention, Technology* (Cengage Learning, 2014).

Her two space history books collected a number of awards including "Best New Mexico Book" (2011 New Mexico Book Awards), Third Place in General Nonfiction (2012 National Federation of Press Women Communications Contest), "Best Book" (2014 New Mexico-Arizona Book Awards), and a Silver Award in Science (2014 IndieFab Book of the Year Awards). She especially treasures the opportunities she has had to speak with icons of space history including Buzz Aldrin, Walter Cunningham, Gene Kranz, Mike Mullane, Harrison Schmitt, and numerous lesser-known but fascinating people.

Loretta is also an award-winning public speaker. She delights in sharing stories of America's space history with audiences from middle school through adults.

Loretta believes in putting her time, effort, and financial support where her interests are. She is an active member in the National Space Society, New Mexico Press Women, the Historical Society of New Mexico, Southwest Writers, and the New Mexico Book Co-op.

Contact Loretta through her website at *authorhall.com/ contactLoretta.html*. She would love to hear from you.

Other books by Loretta Hall

Space Pioneers: In Their Own Words (Rio Grande Books, 2014). Annotated excerpts from oral history interviews of ninety men and women who participated in various space programs from 1945 through the space shuttle era. Associated website at *SpacePioneerWords.com*.

Out of this World: New Mexico's Contributions to Space Travel (Rio Grande Books, 2011). A history of contributions by New Mexico researchers that were crucial to the development of America's space programs from 1930, when Robert Goddard brought his liquid-fuel rocket experiments to the state, through construction of Spaceport America, the world's first purpose-built commercial spaceport. Associated website at *NMSpaceHistory.com*.

Underground Buildings: More than Meets the Eye (Quill Driver Books, 2004). An exploration of the benefits and challenges of earth-sheltered construction explained through more than 100 examples in the United States, primarily civic and commercial buildings. Associated website at *Subsurface-Buildings.com*.

From Skyscrapers to Superdomes: Forces in Balance (Newbridge Educational Publishing, 2005). Supplementary text for middle school students, explaining the basic laws of physics and how they influence the design of buildings.

Arab American Voices (UXL Press, 1999). Twenty primary source documents from speeches, memoirs, poems, novels, and autobiographies presenting the words of Americans with roots in Lebanon, Syria, Palestine, Iraq, Egypt, and other Arab nations, with contextual explanations and topics for discussion.

Arab American Biography (UXL Press, 1999, co-author). Seventy-five alphabetically arranged biographical profiles that provide information about the childhoods, careers, traditions, and other aspects of the lives of noteworthy Americans who can trace their ancestry to one or more nations belonging to the League of Arab States, covering over twenty fields of endeavor.

The author, ready for action in a G-Shock simulator at Spaceport America. Courtesy of Jerry Hall